GW01458958

JAMES COOK

AN OUTSTANDING EXPLORER AND CARTOGRAPHER

THE HISTORY HOUR

THE
HISTORY
HOUR

CONTENTS

PART I

INTRODUCTION

～

James Cook, an 18th century explorer, lived during a time when humankind was yet naïve and inexperienced in terms of understanding their environment. His legacy is quite astounding, given the fact that mankind had only begun to develop instrumentation. The people from Europe had never before seen *Hawaii, Tahiti, Australia, New Zealand, Antarctica*, nor crossed the *Antarctic circle*. He achieved fame by mapping the known world, and his maps are so accurate that many were used right into the 20th century.

～

Cook met and mingled with many different cultures, most of which were those of the *South Pacific Ocean*. He and his men even witnessed a human sacrifice. There were three voyagers in all. The first one ran from 1768 to 1771 and included *Tahiti*, *Australia*, and *New Zealand*. The second voyage went from 1772 to 1775 and was mostly spent around the coast of *Antarctica*. Cook's third and final voyage went from 1776-1779. It entailed the western coast of *North America* and *Alaska* where Cook fruitless sought for the fabled *Northwest Passage*. *James Cook* was murdered by the Hawaiians over a tragic incident that should have never happened.

PART II

JAMES COOK, MASTER'S MATE

«Do just once what others say you can't do, and you will never pay attention to their limitations again.»

James Cook

❧

James Cook looked out the window of the dingy shop he tended in the fishing village of *Staithes* on the Southeastern coast of England. It was stuffy, so he opened the window. The salty breezes came wafting in, and he stared into the distance. What wonders lay beyond the horizon of the sea? The fishermen, draped in brine, came into the harbor to unload their catch. They complained about the narrowness of the channel; the sudden fog that kicked up just before they hit that school of lively codfish. Then they dumped the load from their nets, and the eyes of the shop cat grew wide with envy. Cook could see the clear blue of the sky that fuzzed out just before it met the horizon.

"That's where I belong,"

he thought to himself, as he opened the door for the cat. The sailors hollered at him, but he grinned and shrugged his shoulders.

~

The shop owner, **William Sanderson**, liked **James Cook** as a teenager, but soon realized the boy he was unsuited as a shopkeeper. Instead of waiting on his customers, Cook daydreamed as he gazed out the window, talking about life at sea. *Sanderson* then sent him to some **Quaker** friends of his in the nearby village of **Whitby**, another fishing village that had a hearty coal mining district. Coal was one of the greatest commodities of Great Britain in 1746 when he was employed by **Henry** and **John Walker** as a seaman. He plied the waters to London, the city of Tyne, then on to the Scandinavian countries via the North Sea. The English coast is rocky and windswept, but Cook showed great skill in keeping the ship from foundering. While working for the Walkers, Cook attended school and studied math, trigonometry, and astronomy. That was what seamen learned so they could measure distances and plot their positions by the stars. Then he was transferred to a naval vessel by the curious name of *Freelove*. (It was, after all, a warship!)

~

When he graduated, he worked for the merchant marines

until he was eligible to join the **Royal Navy** where he knew he could move up the ranks. There was a war going on at the time – the **Seven Years' War** (1754-1763) – one of those perennial wars between his native **England** and **France** which often spread to other European countries. Great Britain and France, in particular, were fighting on the continent of America for possession of the lands west of the Ohio River and control of the lucrative trade routes in the West Indies off the Southeastern coast of the North American continent.

"AMBITION LEADS ME"

❧

During the war, Cook's ship transported troops and weapons to *Flanders*. He was then transferred to sail to the *Baltic Sea* with military supplies. In the year 1751, he was on the ship called *Three Brothers* with *Robert Watson*, the first mate. The Walkers, who owned nearly all the ships in the area, hired him for two years to work as the mate on the *Friendship*. During his down times, Cook pored over *John Seller*'s text, *Practical Navigation*. From that book, he learned the use of the basic navigator's instruments of the 18th century – the azimuth compass, the ring-dial, *Davis*'s quadrant, and the cross-staff. The azimuth measures the distance between the horizon calculated between the direction of the sun, another star from *Polaris*, the *North Star*. By making comparisons, ships can determine how far they've traveled and in what direction. The *Davis*'s quadrant measures the altitude (height) of the sun or moon and the cross-staff measures angles to determine the latitude and longitude. Soon afterward, Cook was promoted to command his ship, the *Friendship*. Even though the Walkers sent their sailors across the Atlantic to *Newfoundland*, Cook was never asked. He was much too valuable, having the uncanny ability to keep his ships afloat along the English coast. Most of their sailors had wrecked at least one or two ships because many sections of that coast can be very treacherous.

COOK IN THE SEVEN YEARS' WAR

~

C ook craved a life sailing the world, and – much to the disappointment of the Walkers – he joined the *Royal Navy* in 1755. *Captain Hamar* of the *HMS Eagle* was very pleased to have *James Cook*, as he had so much prior experience. Most of the sailors on the *Eagle* were pressed into service because of the war and had virtually little experience if any. In June of 1755, the ship needed a lot of preparation before their official review at the port of *Portsmouth*. Finally, by August, Hamar was put under the command of the admiral himself, *Admiral Hawke*. After the *Friendship* traveled south, it encountered monstrous gales and brought back in for repairs at *Plymouth*. The displeased admiral then relieved Hamar and replaced him with *Sir Hugh Palliser*.

~

EVEN THOUGH THE *Eagle's* mainsail was broken, the ship

captured a French warship and sunk another. Cook was then given command of the cutter, *Cruizer*, and promoted to the boatswain. As a boatswain, he took care of the sails, ropes, and anchors. He and his men then had an engagement with two French warships and crippled them. As Cook sailed toward **Cape Barfleur**, off **Normandy**, he was met by a fierce cold gale. It was the winter of 1756. Sailing further south, Cook and his sailors were joined by Captain Faulkner of the **HMS Windsor**. There, they were met by two other British ships, and Cook was placed in charge of the **Eagle**. Soon afterward, many of his men were dying – not because of the war – but because they had very little by way of clothing. Those men had been pressed into service and weren't provided with uniforms. Cook had been contacting the admiral in that regard but didn't receive sufficient clothing until months later. Cook himself took ill on this journey and had to take a leave when the ship reached Portsmouth to be refitted.

~

AGAIN AT SEA off the northwestern coast of France, Cook and his two sister ships came across a distant brigantine. It was a dark and rainy night. Once the ship broke the horizon, Cook sighted the French flag over a distant ship and fired at them with his 50 guns. It was a long and bloody battle. Fiery arrows struck his deck, setting the sails and the ship on fire. Seven of his men were killed and two more died before morning. The foremast and the mainmast broke, but Cook limped along, defeated the French vessel, and took many prisoners. Then Cook was forced to sail off until taken into tow by the **HMS Medway**. The two ships then ran into a damaged **East Indian** ship fighting for the French. Without much of a struggle, the captain surrendered his cargo and bribed his way out of custody.

~

ALTHOUGH *HUGH PALLISER* recommended a commission for Cook, neither Palliser nor Cook nor his sponsors, the Walkers, knew any influential nobles or members of the *English Parliament*, so he was passed over for a promotion that rightfully belonged to him.

COOK: MASTER OF THE SOLEBURY

~

I n 1755, Cook completed his studies at *Trinity College*, passed his master's naval examinations and was finally made a master of the *HMS Solebury* which was going to patrol the *Scottish coast* and the *Orkney Islands* off the northern coast of Scotland. When the *Solebury* was in port for a refitting, Cook was made commander of the *HMS Pembroke* slated to sail for *North America*. The British controlled America at that time and were looking to expand their territory westward. However, the French were also in the process of setting up colonies and even had control of *Quebec*, areas of Southeastern Canada and America with the help of the Native Americans. For years, the French fur traders had established relationships with the various tribes in North America, and they became allies during this phase of the *Seven Years' War*, which was also called the "*French and Indian War*."

BATTLE OF FORT DUQUESNE

~

Fort Duquesne was at the junction of two rivers near current-day *Pittsburgh*, and a target for the British, as it lay within the northern area of the *Ohio River Valley*. *General Braddock* was in charge of an attack on the fort. *George Washington* and *Daniel Boone* were both serving under him during that battle. Against the advice of both Washington and Boone, Braddock insisted on using the traditional block formations for the English troops. However, the Frenchmen, the Americans, and the North American tribes fought from behind rocks and bushes like the backwoodsmen they were. At this, the Battle of Fort Duquesne, the English were roundly defeated. Braddock himself was shot and later died.

~

LORD JOHN LOUDOUN was then in charge of British North America. He and *General William Pitt* were ordered to plan the strategy to seize the fortifications at Quebec and Louis-burg, near Cape Breton Island in the Gulf of St. Lawrence. Great Britain sorely needed naval vessels and *James Cook* on the *HMS Pembroke* was one of them. *Samuel Holland*, a midshipman, was an expert draftsman and Cook took an interest in the subject. While they were crossing the North Atlantic, Cook learned draftsmanship from him.

~

IN 1756, when Cook and the *Pembroke* arrived at the *Bay of Gaspe*, off the coast near Quebec, he and his men destroyed the French fort at Louisburg and a number of the French fishing villages there. That action deprived the French soldiers of provisions eliminating any future threat. While there, Cook drafted a complex map of the Gaspe Peninsula which was extremely accurate. It was published as a sample of an expert's work and displayed at Trinity College.

~

MANY OF THE charts they had before this time were grossly inaccurate so Cook was called upon to survey and draft charts of the St. Lawrence River, Newfoundland, Nova Scotia, eastern Labrador, the eastern area of the province of Quebec, St. John's Island and the northernmost coast of North America. Cook not only used the draftsman's tools but plotted out the latitudes and longitudes of the area with his navigator's tools using his knowledge of astronomy and trigonometry. Cook had become a highly skilled cartogra-pher, and his abilities were tapped by ships' captains to plan naval attacks during the *Seven Years' War* and after that as well. While sailing the coast of Newfoundland, Cook chartered the shores so well that his maps were used for two-hundred years!

~

JAMES COOK WAS ASKED to come aboard a naval transport vessel on the way to Quebec. The captain of that vessel, **Captain Knox**, asked him to guide him and the British fleet through the waters of the St. Lawrence River and point him away from dangerous areas. In his logs, Captain Knox recorded that Cook

> *"...pointed out the channel to me as we passed,*
> *showing me, by the ripple and color of the*
> *water where there was any danger; and*
> *distinguishing the places where there were*
> *ledges of (unseen) rocks, sand bars, and mud."*

As a result of his expertise, every single English ship arrived at Quebec whole and undamaged.

THE BATTLE OF QUEBEC

~

The marines disembarked and climbed the formidable cliffs overlooking the city when they arrived, while Cook and the other seaman in the British ships pommeled the shore with their gun batteries. Cook also disembarked and fought in hand-to-hand combat on the Plains of Abraham, so named after the man who owned the farm on that land. Eventually, Quebec ran out of supplies and provisions as well and surrendered to the British on September 13, 1759. There were more actions in the *Seven Years' War* in the West Indies, and across the Atlantic in Prussia. After several battles, many of Cook's men were ill. Illness often inflicted soldiers during wartime. Because they lived in such close quarters and little was known about sanitation, disease spread. In 1763, the *Treaty of Paris* was signed, and the disease experienced by both sides was one reason for ending the war. By the treaty, Great Britain gained control of Spanish Florida and Northern Canada where fishing became a lucrative industry.

∽

AFTER THE WAR, *James Cook* visited Essex where he met the second love of his life, *Elizabeth Batts*. His first love, of course, was the sea. Elizabeth was very personable and a great conversationalist. Her family was a humble one. Elizabeth's first husband had died, and her mother was the daughter of a currier. She knew of Cook's fame as a sailor and knew he would only come home when in port. Nevertheless, she was an independent lady, so she was willing to accept him on those terms.

THE FIRST VOYAGE – PART ONE

～

After the war, the *Royal Society* of Great Britain hired *James Cook* for a secret plan, which they disguised as a mission to plot the phases of *Venus* and observe *Jupiter* and *Mercury*. It was their real intention to discover what riches lay beyond their lands. The astronomical portion of their voyage was the assignment to make telescopic observations of Jupiter, Mercury and measure the phases of Venus. The phases of Venus were utilized by *Nicklaus Copernicus* as one of the proofs that the earth revolves around the sun – the heliocentric theory, which was published in 1543. The phases of this planet have been the subject of scientific inquiry for years, so their cover-up worked well.

～

THE EUROPEAN SAILORS and tradesmen who traveled the *Mediterranean Sea* in those days had seen dark foreigners in clothes of silk with golden trimmings. They had seen camels and heard for years grand tales of sea monsters churning up the waters of the great sea (Atlantic Ocean) that lay to the west. They were told that the waters at the equator boil in huge bubbles. They had seen fish so large it took many men to carry them ashore. The Spanish and Dutch explorers had gone to sea and established trade routes that could make poor men rich, and the English too wanted this wealth. They heard stories about how the Spanish explorers had traveled a huge distance to a land they called *California*. And then there was South America, which most British people had never seen. There were many mysterious islands in the Pacific Ocean. In the *Bible*, they read the story of the port of *Ophir* from which *King Solomon* obtained his wealth of silver, gold, pearls, and ivory.

~

THE BRITISH IN particular also had a secret desire to find the "*Northwest Passage*," a waterway that was said to lead from the Atlantic to the Pacific Oceans. Many good seamen were lost to the fury of the waters at the tip of Africa and of South America that such a waterway could save lives and cut the length of the journey from Europe to China. The *Royal Society* planned on doing that in the future, if Cook's initial voyages were successful.

~

IN 1769, *James Cook* left Portsmouth in his trusty barque, the *Endeavor*. A barque was a three-masted vessel of great height designed to collect the great winds. The mizzen masts were triangular, and the rigged mast in the aft assisted the work of the rudder. At the *Cape Horn*, at the tip of the African Continent, the great waters of the two oceans meet. On a fine day, the voyage around *Cape Horn* is pleasant, but on a windy day, a ship could crash into the rocks be drawn in by the fierce currents. When Cook traveled around *Cape Horn*, a white bird with huge wings glided overhead. It was called the *Albatross* and had the largest wingspan of any known bird.

∾

COOK KNEW THE SEA; he knew the longitudes and the latitudes. He knew the old seafarer's proverb:

> *"Below 40° South, there is no law, below 50°*
> *South, there is no God."*

That area is at that line of demarcation where the waters explode. The winds grow wild there, and one can hear the call of the "*sirens*" according to the old myth. "*Sirens*" are mythical females who wait on the rocky prominences calling for the sailors to join them. If they do, though, it is said that they are enchanted and never return from the sea.

CAPE OF GOOD HOPE

∿

From *Cape Horn*, Cook sailed toward the tip of South America, called the *Cape of Good Hope*. To avoid going around the Cape itself, Cook sailed through the *Strait of Magellan* and the *Tierra del Fuego*, which cuts off the southernmost tip of the continent. From the *Endeavor,* he looked out at the destitute natives who wore few clothes. Their homes were made of grass and sticks. However, they seemed to want nothing more unless offered bread, which they loved. From there he planned to take measurements of Venus and then veer west toward the island of Tahiti in the South Pacific.

FRENCH POLYNESIA

~

There, in the middle of the South Pacific Ocean, they came upon some delightful islands which were adorned by wild tall palm trees waving in the trade winds. This collection of islands they called *Otaheite* (today's *French Polynesia*). Each sand strewn island was given names after their physical characteristics – *Lagoon Island*, *Bow Island*, the *Groups*, *Thrumb Cap*, *Tahiti*, *Chain Island*, and *Bird Island* named after the flocks of white herring gulls that swarmed around the trees, picking at the coconuts. This was a virtual paradise by comparison to the dull hills of the Terra del Fuego along the Strait of Magellan.

~

THEY DROPPED anchor at *Royal Bay*, which the natives called *Mataria*. After setting his ship's guns in place in case of trouble, Cook took a landing party with him, consisting of *Mr. Banks* with his spyglass, *Dr. Solander*, the botanist, and *Mr.*

Green, the astronomer. The natives weren't ignorant people, having been visited before by trade ships. Once the members of the party started exploring, one insolent lad grabbed Mr. Green's musket and fired it. Without delay, Cook had his men discharge their guns. This they followed up with gunfire from their ship, aimed over their heads. On the next day, curious unarmed natives waited patiently ashore as Cook and company approached in a friendlier manner. They had trinkets with them for trade. For those, they were given coconuts and round green fruit.

～

COOK and his men then erected a small fortress, and Cook set up his tent and equipment for astronomical observation of the planet Jupiter as he had been instructed. The natives even assisted them, tightening the poles to hold the instruments in the deep sand. A native came forward by the name of *Owhaw* and mimicked gestures indicating communication. They understood.

～

DURING THEIR VISIT, there was an incident. The cook aboard the *Endeavor* threatened the wife of the tribal chief, *Tuvourai Tomaida*, demanding a stone for his hatchet. She was frightened and refused. In front of the natives, Cook had his men take a lash to the offensive cook. The natives cried out for mercy upon his second stroke. No mercy was granted, and the natives cried out in sympathy for the punished man.

～

WHEN MR. GREEN went searching through his tent to find his quadrant for measuring Jupiter's position, he found it had been stolen. Mr. Banks then went into the underbrush and courageously interceded with them, trying to tell them it wasn't a weapon. Triumphantly he returned with the thief. The thief was *Tootahah*, the chief of the minor tribe. After the exchange of some primitive sign language, the quadrant was returned.

~

ONCE THE MATTER WAS RESOLVED, the natives exchanged breadfruit and coconuts for nails, which was a favored trinket they used for necklaces. Breadfruit grew plentifully on a tree related to the *Mulberry*, and there were a lot of those trees in the village. Breadfruit is a starchy vegetable, suitable for eating and tastes like potatoes. After being invited to visit the ship, *Chief Tootahah*, *Tuvorai Tomaida*, and two others became friends. *Mr. Gore*, Green's assistant, and a few natives took a longboat to another unnamed island to set up their instruments and observe the phases of Venus. The weather was crystal clear, and all their readings were excellent, especially those made with their telescope.

~

DURING THEIR STAY, Cook and his men celebrated *King George III*'s birthday and feasted with the tribal people, who played musical instruments for them – flutes and hand-made drums. They also sang in their native tongue. While the natives seemed gentler at this point, they were notorious thieves, taking anything that looked glittery or interesting. Equipment was missing, so Cook ordered his men to seize the native canoes that were loaded with fish, and tried to ransom them for the return of the equipment. When that didn't work, he had the thieves confined. Being in captivity was something they feared very much.

~

WHEN THE SAILORS collected stones and heavy objects for ballast, they ran out of them and started collecting the bones of the dead from their burial mounds. An incident arose, which Mr. Banks was quick to correct by forbidding the men to use the bones of the dead.

~

IN THE MEANTIME, Dr. Solander and Mr. Banks collected watermelon seeds, oranges, limes, lemons, and other plantings. Some of those they planted for the natives and took the rest with them. As they were planning for their departure, another problem arose. Two of Cook's crewmen, *Clement Webb* and *Samuel Gibson*, went missing. It was later discovered that they had been kidnapped by the native women who claimed to be in love with them. Cook retaliated by grabbing *Tubourai Tomaide* and some other natives and took them aboard the *Endeavor*. Cook then told them the hostages would be returned if his men were returned. *Chief Tootahah* then spoke to Cook and led him back to where the prisoners were held. Upon the chief's orders, the sailors were released. With him, *Tootahah* brought a thirteen-year-old boy, *Tupia*, whom he offered as a guide to help Cook explore the other islands in the South Pacific. Graciously, Cook accepted, and the delighted lad accompanied him. As they left, Cook and the sailors left gifts of white linen, spyglasses, nails and ax heads. Tupia said his prayer to the god of the wind who was called *Tane*, and they proceeded.

∼

It was now July of 1769. Their next destinations were the *Society Islands*, Northwest of Tahiti and French Polynesia. *Huaheine* was the largest of the Society Islands. Upon reaching Huaheine, the Islanders were afraid until they saw the brown-golden face of Tupia at the bow of the first canoe. Upon disembarking, they were greeted with friendliness. Their traditional greeting was the sharing of everyone's names. *Oree* was the chief of the tribe, and they called Cook "*Cookee*." Every one of the smaller islands had the same kind of greeting. The sailors exchanged golden coins and a colorful plate for a small hog.

∼

After they boarded their ship, Tupia warned them not to visit the next island, *Bolabola*, because the people were hostile. Despite that, Cook, Mr. Banks, Dr. Solander and the others disembarked anyway. They proceeded to make the same greeting as they did on Huaheine. Surprisingly, it worked, and there was no hostility. The chief of *Huaheine* was named *Opoony*. Cook and his crew expected to see a tall gallant dark-skinned man, but he was feeble and half-blind, no doubt from having warred with some of the other local tribes. After that, the crew hoisted the English flag and claimed possession of Huaheine, Otaha, and Bolabola for the English crown. Then they traded with the Islanders and surveyed the islands. They weren't as picturesque as the Society Islands but were sandy and peaceful and full of fruit. As they were leaving, it was low tide, and they noted some coral cliffs buried in the sea, but the ship was undamaged. The waters were shallow in many places, and navigation was very tricky. Then they returned to French Polynesia, landing on the opposite side which they hadn't seen before. Then, above one of the larger islands, *Oheteroa*, a comet passed over. The indigent people felt that the comet was an omen that the cruel warriors from Bolabola would attack them. Cook and his men, however, made observations of the comet and measured the direction and length of the comet's tail. From there, Cook turned south.

PART III

FIRST VOYAGE: AUSTRALIA AND NEW ZEALAND

«Ambition leads me not only farther than any other man has been before me, but as far as I think it possible for man to go.»

James Cook

THE LAND "DOWN UNDER"

⌇

They then proceeded to the *Great Barrier Reef* off the coast of Australia, which Cook called "*New Holland*." Its blue-green waters are clear enough for one to see the pink, blue, and red coral of the long chain. Rising from the Great Barrier Reef is the coral island called *Great Keppel Island*. It flourished with life. The great kookaburras and rainbow-colored lorikeets fly from tree to tree enjoyed its foliage and fruit. A lorikeet resembles a parakeet, except for the fact that it is much more colorful with red, blue and yellow feathers. A kookaburra looks like a small kingfisher, but lives in the trees. it emits a laughing sound. There is an amusing rhyme sung by the children and even by adults regarding this curious bird:

> "Kookaburra sits on the old gum tree,
> Merry, merry king of the bush is he.
> Laugh, Kookaburra, laugh, Kookaburra,
> Gay your life must be!"

The water gum tree is very plentiful in Australia and New Zealand. It can grow as high as fifteen feet tall. The kookaburra enjoys the nectar of the yellow flowers that grow upon it, and so do the bees. Most of the gum trees grow along the coastline.

∾

THE BEACH IS full of white sand, bleached by the sun. Cook told his men that the island was claimed by **Captain Wallis** of the Royal Navy.

∾

FURTHER SOUTH, along the eastern coast of Australia, the **Endeavor** came upon a small bay loaded with giant stingrays, and he called it "**Stingray Bay.**" It was nearly circular and sheltered. Mr. Banks and Dr. Solander disembarked there and collected botanical specimens. Later on, Cook changed the name of this to "**Botany Bay.**" In the year 1788, Great Britain established a penal colony there but was forced to move it to what is today the city of Port Jackson. The bay itself is too shallow for larger ships; the adjoining lands were full of swamps, and the soil is too poor for farming. Even buildings that were constructed there collapsed in a short period of time.

∾

FURTHER INLAND on the continental shelf, there were many tribes of dark-skinned people with high cheekbones. Genetically, these people were a mixture of the Southern races. They are the **Aborigines**. Their culture is mysterious, but they carved many figures in the many caves and nearby rocks representing their lives. They saw themselves as partly of nature and partly human. They were very attached to the areas where they lived, and they believed that the land there sustained them. The Aborigines migrated into various areas of Australia depending upon the seasons. Descendants of many of the Aborigines live in Australia today along with the Englishmen who settled there later on. In the early days, the Aborigines raised yams, hunted game and fished. Mr. Banks noticed a fish that had been gutted and was astonished that it appeared to weigh hundreds of pounds. One of the species of a game they came across resembled a small turkey. It is called a "*bustard*," and wandered through the dry grasslands and low-lying woodland eating insects and small rodents. It can weigh up to twenty pounds. In the abbreviated bays that were along the coast, oysters were found in the mud, and the men had hearty meals of these oysters. If friendly relations could be established with these people, no doubt, many pearls could be harvested. The large bay alongside the grasslands was full of loam and sand.

$\sim$

IT WAS difficult to create friendly relations with these people, as they were both suspicious and fearful. Some men even came rushing out of small hovels with long spears and shields made of bark. They heaved the lances at Cook and the men, so the crew was forced to fire a few muskets in the air to dissuade them. Cook left presents for the Aborigines, even including nice cloth and vases. The natives, however, never took possession of the gifts, and the men found them still there the next morning. To avoid further conflict, the men explored further inland and south, where there were no villages. The trees were magnificent South of Botany Bay where the soil was much more fertile. The short trees there were full of the rainbow-colored lorikeets and even cockatoos with their proud crests of feathers.

∾

WHEN MR. BANKS, Dr. Solander, and *Mr. Monkhouse*, the ship's surgeon (doctor), explored further inland, they discovered huge clouds of pesky mosquitoes. The land that lay there along the Eastern coast and somewhat inland was and is still called "*New South Wales*."

∾

AT A DISTANCE SOUTHWEST of Keppel's Island, the Cook crew sighted a magnificent land full of low-lying foliage and hemmed behind with tall mountains. They thought this land was "*terra australis incognita*." In his journal, Cook described it as:

> "*The land on the sea-coast is high with steep cliffs*
> *and the back inland has very high mountains;*
> *the face of the country is of a hilly surface*
> *and appears to be clothed with woods and*
> *verdure*." (green foliage)

This wasn't the legendary "*terra australis*," though; this huge island east of Australia called New Zealand, and Cook realized that after further exploration. This land mass wasn't a part of Australia either because it is geologically different from Australia and believed to have been separately formed.

NEW ZEALAND

❧

New Zealand is split into two distinct islands – one smaller and another long and narrow. At the head of the long island was a delightful bay they named *Queen Charlotte's Bay* after *King George III*'s lovely wife.

❧

AS THEY NEARED THE LAND, a great many indigents came rushing out from the woods with long spears. The sailors who were manning the lookouts on the ship gave the sailors sufficient warning, so they backed out of the harbor. As they were moving out, the native people splashed into the water around the *Endeavor* and heaved heavy lances at them. Then Cook responded by firing a few muskets over their heads. Astonishingly the natives ignored them! Then Cook and his men were attacked, and one sailor shot one of the indigents dead. After hearing the loud crack of the rifle, the natives froze in their tracks and stood motionless, staring at their fallen comrade. Then they stopped throwing their lances, and Cook retreated just a short distance. The tribesmen talked with each other loudly, apparently trying to understand the event. Then the courageous Tupia climbed off the ship, and cautiously approached them. He spoke in the language of one of the islands in the South Pacific called *Oheteroa*. Although Tupia's dialect was somewhat different, it became clear that they understood him. He told them the sailors wanted to trade for iron – as they knew that there must be much of it in the mountains. The warriors felt that the iron was of little value, so they let these strange white men dig in the bogs for iron. All the sailors could find to trade with them were long feathers, which they seemed to prize. None of their other trinkets or gifts interested them. Because these indigent people seemed to want very little, Cook called the bay where he landed "*Poverty Bay*." Poverty Bay is on the Northeastern coast of New Zealand. Tupia kept up his caution and reminded the tribesmen that they would be killed if they showed any signs of violence.

∾

THERE WAS a thin river inlet nearby, but the men discovered that it was brackish and undrinkable. As they moved along toward the sea, they came upon huge waves that rose many feet in the air and crashed upon the beach. Cook's men were shocked; however, then they saw the natives go out upon the great waves in their little canoes and maneuver them as if they were surfboards.

~

THEN, without warning, one of the villagers attacked unexpectedly and kidnapped one of Cook's sailors hauling him into a canoe. In response, the sailors had to fire their muskets, killing the perpetrator. The kidnapped sailor, Tayeto, jumped into the sea and swam back to the *Endeavor*. Cook then named the bay just south of there "***Kidnapper's Bay***." Rather than attempt another contact, he thought it wise to set sail, and moved south, dropping anchor slightly offshore. There he and the men set up their equipment and measured the transit of the planet Mercury.

~

IN SOUTHERN NEW ZEALAND, the natives, the Maori, were apprehensive at first. However, the other people from the indigent population were friendly, and Cook's men traded some of their goods for fresh fish. At the southwest end of New Zealand, they discovered what they called "***Dusky Bay***." The birds were colorful, the seas were replete with fish, and there was a small game nearby, so the men feasted. By then, Cook had carefully chartered the entire perimeter of New Zealand, and his draft showed great accuracy even by today's standards. As they pulled away from the shore, a gentle breeze came up. However, it became stronger and stronger, and the *Endeavor* was tossed about hither and to in the wild surf.

~

IN ADDITION TO NEW ZEALAND, they sailed around the

nearby island, currently called *Tasmania*. There they saw huge numbers of birds. Among them were robin-like birds, tiny wrens with blue cheeks, gray birds with bright red tails, colorful small birds and brightly-colored parrots. After midnight, nightingales burst into song. In the bays that led out to the sea, there were many pelicans who swooped down into the waters and arose with great fish which they took to their nests in the grasses. Cook's men then dropped baited lines into the water. Great blue crabs snatched the bait, and the men feasted upon them.

~

THEY WENT ASHORE and met a tribal family, who was very gracious. However, as they sat with the family over the fire, Cook's men were horrified because they saw remnants of human flesh and bones. According to their research, the early tribes of *Tasmania* consumed the human flesh of enemy tribes. The sailors treated them cautiously but traded with them and the neighboring families. Like many of the people in that region, they prized nails, which they used for building modest dwellings made of wood and thatch.

~

IN BOTH AUSTRALIA and New Zealand, Cook especially noticed the health of these people. There were very old people as well as those who were younger. The elderly men were often bald and had lost their teeth, but – other than that – they displayed no other symptoms of the disease. None had any sign of illness or deformity. Their skin had no rashes, and the young and middle-aged people were very strong. If they were cut or wounded in the course of their daily chores, Cook noted that they healed very rapidly. Over the years, this area was visited by the French, Spanish, Dutch, and other Englishmen. A few of the people had been hit by musket fire but showed no signs of lingering infections. They were naked and showed no shame about it. Tupia was able to communicate with the people of New Zealand, but the language of the *Australian Aborigini*'s was foreign to him.

～

THEIR NEXT HEADING was toward *New Guinea*. From there, they planned on crossing the Indian Ocean, going around the Cape of Good Hope at the tip of Africa, then along the western African coast and north to England.

～

BY THE 23RD OF AUGUST, 1770 they were on their way to the island of New Guinea. There they came upon a dangerous shoal and then many other shoals. In observing the island with their spyglasses, the men became envious of the fruit there – coconuts and plantains (like small bananas) but didn't tally because the people of New Guinea were somewhat hostile and shot flaming arrows at the sailors. From there they sailed south rather than further antagonize the tribes of New Guinea.

～

WHILE CROSSING THE INDIAN OCEAN, they were far enough south to observe the *Aurora Borealis*, shooting up from the horizon in its green and blue colors as the sun was setting on the southern continent of Antarctica. There were islands strewn around there that had been settled by the Dutch. From the *Endeavor*, they became nostalgic when they saw flocks of sheep grazing upon the hills of **Timor** and the island of **Savu**, off the coast of Indonesia. There they stopped to get supplies — those included buffaloes, sheep, hogs, eggs, coconuts, limes and palm syrup. Savu was headed by a king, and the people were Christians. From there they diverted to **Batvia**, a city in the Dutch East Indies. At Batvia, there was a wonderful greeting ceremony held by the governor in their honor. There were also European-style hotels, and the men tried to enjoy a few nights. While there, Cook discovered that the *Endeavor* was in great need of repair especially upon its hull. The worms had worked their way into the timbers of the ship's bottom and needed reinforcement. They worked hard and rapidly to refit the ship.

～

THE MEN HAD BEEN RELATIVELY healthy until they reached the **Dutch East Indies**, but Tupia's companion, **Taketa**, became extremely ill and died. So, too, Mr. Banks and Dr. Solander became deathly ill. Slowly they were nursed back to health by Mr. Monkhouse, the ship's doctor. When Cook was hit with the same illness, but fortunately it wasn't too serious. The disease had spread to the Dutch East Indies and resulted in several deaths. Cook's dear translator and friend, Tupia also died. Cook was very upset about the loss and became extremely concerned about the health of his men. He took as many precautions as he could to provide fresh water and food to the ill. Desirous of escaping the East Indies, he and the able seamen left as soon as possible, heading westward toward the Cape of Good Hope and Capetown South Africa. It was now March of the year 1771.

∼

WHEN THEY REACHED the municipality of St. Helena in South Africa, Cook was horrified at the treatment of the people there. There was a colony of African slaves owned by the British who headed Capetown. Cook was embarrassed by the inhumanity of his countrymen toward those slaves who were forced into hard labor. Rather than provide those men with carts which were readily available, the slaves were forced to carry provisions piled high on their heads and made to trudge through the rough country.

∼

BY THE TIME Cook made his next voyage, though, the conditions for the slaves were mercifully improved. Cook had made his opinions known before his departure and word reached England through his logs which were turned in here and passed along to his home country.

∼

THE VOYAGE up the western coast of Africa was a familiar one for him. In July 1771, he and the remainder of his crew arrived in Plymouth England. The Royal Society who had commissioned him to observe the planets – specifically the phases of Venus – were delighted with his reports. The members of the Royal Society and the Admiralty were also extremely pleased with all of his charts.

∿

HIS WORK WAS PUBLISHED in 1773. Although the press criticized the editor, *John Hawkesworth*, for having eliminated a lot of information, the material contained in the 3-volume series was generally well-received. The title of the book was called *A Journal of the Voyage around the World.*

THE MYSTERY OF "TERRA AUSTRALIS INCOGNITA"

～

C ook and the astronomer, Mr. Banks, had long ago heard that "*Terra Australis incognita*" was a great land on the Southernmost base of the earth. In 1570, *Abraham Ortelius*, a cartographer, had theorized the existence of such land at the bottom of the Earth. This mysterious land was assigned to Cook for his second voyage. What Cook didn't know was whether or not Australia connected to this theorized southern land mass because he hadn't circumnavigated Australia at that time.

THE PREPARATION FOR THE SECOND
VOYAGE

≈

I n August of 1771, by order of *King George III* of England, Cook was appointed the head of his majesty's navy. It was an honorary title, however, as he hadn't gone through the various ranks of the navy according to the *Earl of Sandwich*, head of the *Admiralty*. He also received honors by the *Royal Society* for his astronomical observations and charting of the islands never seen before.

≈

FOR THE SECOND VOYAGE, two ships were commissioned – the *Resolution* and the *Adventure*. Cook headed the larger ship, the *Resolution*, which carried 112 men and *Mr. Tobias Furneaux* commanded the *Adventure*. It had a crew of 80 men. *William Hodges* was employed as a landscape painter who would be useful in accompanying the written descriptions provided by the logs of the *Endeavor*. *William Wales* and **William Bayley** were the astronomers, and **John Reinhold** along with his son were the historians.

～

THE *EARL OF SANDWICH* was on hand for the departure as was **Sir Hugh Palliser** under whom Cook had served in the merchant marine services. After hearing about the gales and rain Cook encountered when crossing the great seas, Palliser advised Cook to keep a controlled fire between decks to the area would be warm and dry for the men. On July 13, 1772, they set sail from Plymouth.

PART IV

SECOND VOYAGE: ANTARCTICA AND THE SEARCH FOR THE SOUTH POLE

«Gifted women musicians and composers rarely received their due.»

James Cook

~

On this, his second voyage, Cook and his crew came upon a most interesting phenomenon. It was observed by the astronomers as well as the curious sailors – a partial eclipse of the moon. Their first landing was at the Cape of Good Hope, where they collected more provisions and plenty of drinking water. As they knew, the weather would be cold; each man was furnished with a warm jacket and thick trousers. There was a brisk south wind when they left, and later there were stiff cold gales, replete with rain and hailstones. Because of the cold, they lost some of their livestock. It was now November and moved into December quickly. As a precaution, Cook had a supply of liquor and spirits to warm the men, for which they were grateful. On the second week of December, the ships ran into sheets of ice atop the seas. A low-lying glacier was found after the haze of the evaporation had lifted. It was huge, according to Cook's report, rising as high as fifty feet. There were many more glaciers rising from the ocean waters, and one of them measured two miles long. They had to navigate around many "*ice islands*," with sharp edges and mounds of ice appearing above the water line. Some of the ice floated like rocks, and they had to be careful of collisions. Thick fogs descended upon them all around. Cook had caps made for the men and had their sleeves lengthened. In the *Southern Hemisphere*, the seasons run opposite those up north, and this was supposed to be summer in the southern climates. Of course, it was so far south and so close to the South Pole that it was frigid. They were surrounded in January by a huge ice field, and it took a while before the two ships could break the ice into chunks to allow passage. Cook had wanted to go even further south, but the ice kept impeding his path, so he sailed westward looking for an icy inlet.

~

As they sailed, they spotted great penguins waddling along peaceably. The huge albatross birds they had seen in Australia showed up here as well. By mid-January, they could proceed no further south. Cook was disappointed as he had read that the French had gone further south. By the next month, however, the *Adventure* couldn't be sighted, and signal fires had no return responses. Cook and the *Resolution* then had to proceed alone. However, they had prepared for that eventuality by agreeing to meet again at *Queen Char-*

lotte's Bay in New Zealand where they planned to stop on the way back.

~

One night about midnight, they observed what they called the "*Aurora Australis*," after the theoretical name given the area *(terra australis incognito)*. Unlike the Aurora Borealis they'd seen earlier, the colored lights were extremely strong and gave off rays that curved and circled in magnificent shades. The glaciers were extremely high and could only be seen when the haze had cleared, so they had to be careful to avoid being trapped. From the distance when the weather cleared, they noted huge wide and white caverns penetrating some of the glaciers.

THE DESOLATION ISLANDS

~

A lso known as The *Kerguelen Islands*, the *Desolation Islands* are plateau-shaped groups of islands below the Antarctic Circle. They are, unexpectedly, igneous rocks, formed when there once were active volcanoes in the region. During eruptions that occurred many years ago, molten rock had risen from the magma layer of the inner earth. Over many thousands of years, it solidified and crystallized. Outcroppings it created rose as if in spires and spikes. There were also smooth ice-hardened surfaces that rose from the sea in miles-long slopes, impossible to climb. On the narrow flatter sections of this huge land mass, which is mostly submerged in the Southern Indian Ocean, more noisy penguins waddled along. Thin river inlets wind their way toward the majestic heights of the many snow-capped mountains that were strewn among the islands. The earth there bears a dark ruddy color. This surface, which creeps into the ocean is mostly basalt, that was formed near the surface of a centuries' old lava flow.

Toward the Southeast, is a huge wide frozen glacier later called "*Captain Cook's Ice Cap*." On the high ground lies the main island, *Grande Terre*, measuring about 25,000 square miles. It is a dangerous area to travel, as the men discovered because the winds are as strong as hurricanes. However, there is still enough hot magma locked inside the earth below to leave it ice-free. This is an area of apparent geological contradictions. All the islands here form an archipelago. Cook confirmed that these islands were first visited and claimed for the French in 1772 by *Yves-Joseph de Kerguelen-Trenarec*.

NEAR MISS: BOUVET ISLAND

❧

Bouvet Island, located just north of the ice pack and glaciers is the most remote place on earth. 93% of it is covered with glaciers many thousands of years old. It was a volcano, now imprisoned at the base of the earth and locked there in icy silence. Cook explored some of the areas around it, but a wicked gale loaded with hail and rain, blew them off course to the east and he missed Bouvet's fabled *Cape Circumcision*, an elevated flat plain. All around them were ice islands rising as high as sixty feet in the air. If the waves during a wild storm were to crash over those islands, they could all be smashed into the sea. It was densely foggy there, and sometimes the men thought they saw a passageway, but that only proved to be a mirage or illusion. Fresh water was gleaned by taking ice chunks and melting them. *James Cook* then wisely decided he couldn't find the *South Pole*, being as the entire area was locked in by ice islands and huge long sheets of thick ice. He then proceeded northeast toward the African coast and east of

Madagascar to locate the **Island of Mauritius**. It was not to be found, as the winds whipped them eastward. After the skies cleared the whole black sky again lit up in nature's light show, the **Aurora Australis**, which is the western portion of the aurora borealis. The light show grew and grew until the whole sky was filled with brilliant color. In horror, they noted that all the ice islands they had seen before had split into sharpened pieces, too unsafe to navigate around as they had done before. It was frigid now, and some of the men, including Cook had frostbite on their fingers. The climate was changing for the worse, so he and his men headed for New Zealand once again. They found their haven there in **Dusky Bay** by the fiords near **Resolution Island**, **Anchor Island**, **Long Island** and **Cooper Island**. It was raining when they arrived causing cascading waterfalls. In the waters, the dolphins played, and the seals called out to their mates. Cook and his men searched for friendly natives there but were none were found. The tribal people who lived in this harsh land often warred with each other over trivia, and it was suspected that they might have been killed for the presents Cook had left for them near Dusky Bay.

~

COOK and his crew spent nearly two months in New Zealand, as the area was replete with wildlife. He and his astronomers set up an observatory, an iron forge, and tents for sailmakers. Out to sea, the Humpback Whales were leaped in and out of the waves feasting on the plankton. From the leaves of a black spruce tree, Cook's men created a liquor which they sweetened with molasses.

~

LITTLE BY LITTLE, some indigent tribesmen finally came over. Cook offered them his usual presents in which they showed little interest. However, when he took his bagpipes and drums out, there was song and dance, and the people loved it. Then something unusual happened. Their chief presented Cook with a ceremonial cloth and a hatchet. He then seemed to bless Cook's ship, the **Resolution**, rubbing it with a slick green leaf.

~

COOK and his men hunted seals not only for eating but to provide oil for their lamps, and skins as coverings for their compartments. For his gift to the area, Cook left a flock of geese well-suited for the colder climates and hoped they could reproduce there.

~

THERE WAS little that was unpleasant about the area, except for the black sand flies that emerged when the weather turns rainy. Although the affliction does pass, men were sometimes ill from the insect bites. The air was healthy, however, as was the food, and they quickly recovered.

~

NEXT, they sailed for Queen Charlotte's Bay, hoping to meet up with their sister ship, *Adventure*. As soon as they rounded the entrance, a loud cheer went up. There sat the *Adventure*! They partied and celebrated, and Captain Furneaux displayed the beautiful garden that had grown from the seeds he planted when they first landed there – parsnips, carrots, and potatoes.

~

MOST OF THE men at that point needed the vegetables and fresh fruits. They had symptoms of the scurvy, which is caused by a diet without fresh produce. After the ill had recovered sufficiently, Cook searched for *Pitcairn Island*. However, it so happened that it lay well to the west of them, and they never reached it. Instruments in those days were somewhat inaccurate.

~

FROM THERE, they parted company, and the *Resolution* left first.

THE GRASS COVE MASSACRE

~

While the *Adventure* lingered in New Zealand, Captain Furneaux sent out his botanist and some seamen to collect foliage samples and seeds to take with them for analysis. They left in the small cutter, a ship that was small enough to travel through the winding paths of the river inlets. However, when the men who traveled on the cutter never returned, the ship's Lieutenant sent out a search party to look for them. It was horrifying what they found. Inside a native canoe, the men discovered the remnants of a roasted meal – with portions of human feet and bones still attached to some of the chewed bones! A small mass grave was found nearby for the uneaten parts of the human bodies. Upon seeing that, the men ran away in horror and reported back to the ship. The people who lived there apparently believed that – by consuming the flesh of a powerful enemy –they could eliminate the revenge of the spirits of their ancestors.

~

THE *RESOLUTION* WAS out at sea at that time exploring the South Pacific Ocean. There were countless islands and island chains there. Tonga (known then as the "*Friendly Islands*") had been well known to Dutch traders from the year 1616. This island and the others around it in the South Pacific are now known as Oceania. It's been inhabited by a race of brown-skinned people ruled by a king.

~

IN 1774, *James Cook* and his two ships arrived at *Easter Island*. There was very little foliage on the island, and the weather was hot and humid. The Dutch were the first to discover the island, and there were just several thousand *Polynesians* there at the time. Like *Tonga*, the people who inhabited the island had their chiefs, and the society was divided into a caste system. Easter Island is well-known for its mysterious huge stone statues called *Moai*. They are about 30 feet high when standing. However, there were many inter-tribal conflicts resulting in the toppling of many of the statues. The people there believed that the statues had mystical powers which were lost when the statues were destroyed. The crew found the language to be incomprehensible, except for counting. The land was poor, but the people could raise bananas, sweet potatoes, and sugar cane. Parts of the island looked abandoned, and the civilization was undeveloped.

~

NORFOLK ISLAND HAS A VERY different topography. It has

immensely tall spruce pines and a great many natural flax plants. The timber is hard, and the tree-trunks are very large. About that island, Cook said,

> *"Except for New Zealand, no other island in the*
> *South Sea has wood and mast-timber so*
> *ready at hand."*

Cook named the island after the noble woman, Duchess of Norfolk, Mary Howard, of an old English family lineage. It was uninhabited when Cook visited it.

NEW CALEDONIA WAS AN ARCHIPELAGO. It was first settled in the 17th century before Cook's arrival. The Polynesian people there were called "*Lapita*" and were highly skilled sailors and farmers.

∾

JUST NORTHEAST OF *New Caledonia* was the archipelago called *Vanuatu* or "*New Hebrides*." The people there, the *Melanesians*, were a hardy kind of people related to those in *Fiji* and *New Guinea*. It was first visited by the Portuguese in 1606.

∾

COOK THEN SAILED south toward Antarctica again, as a tip of it protruded into the sea near the Cape Horn. He had been hoping to reach the South Pole from there but was unable to penetrate further south because of solid masses of ice which covered the pole. Their ship was hit with snow and hail-stones, so they had to turn eastward.

~

COOK then and planned to bypass Cape Horn. On their way back, there was a steady gale, but the wind was at their back, helping them travel quickly. They then landed at a bay, and Cook called it "*Christmas Sound*," as they had reached it by Christmas in 1774. There were geese there which they happily killed and used for a feast. In 1775, they came upon an island near the center of the ocean. It had enormous rocks that jutted up from the land. They were later named the *Shag Rocks* and *Black Rock*. The coastal area was strewn with hundreds of emperor penguins and fur seals. Cook was perhaps the second explorer to visit there and named it "*The Isle of Georgia*" after *King George III*. It's now known as South Georgia. A Spanish explorer, *Captain Gregorio* Jerez records having sighted it in 1752, but didn't land there nor lay claim to it. They landed at a huge bay which they called "*Posses-sion Bay*." According to the naturalist, *Georg Forster*,

> "Here Captain Cook displayed the British flag
> and performed the ceremony of taking
> possession of those icy rocks, in the name
> of his Britannic majesty and his heirs
> forever."

In his journals, Cook wrote that the head of the bay

> "*was terminated by perpendicular ice cliffs of
> considerable height. Pieces of it were*

continually breaking off and falling out to
sea."

He further reported that when those huge chunks of ice fell into the water, they sounded like cannons firing.

∼

TO THE NORTHEAST of South Georgia lay a small group of islands which he called "*Clerke Rocks*" after one of his officers, *Charles Clerke*, who had sailed with him aboard the *Resolution*. Along with Shag Rocks, Clerke Rocks has a large population of sea birds called *Cormorants*. They have black wings and white underbellies. Cormorants can dive as deep as 150 feet in search of fish, which is their primary diet. Alongside Clerke Rocks are the South Sandwich Islands, named for the Earl of Sandwich in England. Those islands were later called the "*Hawaiian Islands*."

∼

THE *RESOLUTION* then steered directly toward the Cape of Good Hope at the tip of Africa. The ship needed repair and new rigging and fresh provisions. While on their way there, Cook came upon two ships – one headed up by *Captain Bosch* and another, called the *East Indiaman*, commanded by *Captain Broadly*. Cook asked for supplies, and they were offered sugar, tea, and other provisions, including old newspapers. The men hadn't had a word about home in many months and enjoyed them.

∼

AFTER LANDING at the Cape of Good Hope, the *Resolution* was repaired. In the meantime, Cook scoured the papers for word about his sister ship, the *Adventure*, but found nothing of significance. The *Adventure* had already completed its phase of the mission at any rate.

∽

AFTER REPAIRS HAD BEEN COMPLETED on the *Resolution* and he again stocked up with food and supplies, then turned north along the western coast of Africa, stopped at *St. Helena*, an island owned by the *British East India Company* and headed for home. In July of 1775, the *Resolution* docked in Portsmouth.

∽

UPON HIS ARRIVAL, *James Cook* discovered that his fame was worldwide. He was promoted to Captain and made an official officer of the Greenwich Hospital. ("*Greenwich Hospital*" refers to the naval college as opposed to a medical facility.) Cook was also given an honorary retirement from the Royal Navy. However, he agreed with the *Admiralty* to be permitted to continue with his voyages and receive assignments from the Navy. He, having been hired by the Royal Society to make his astronomical observations, was awarded a fellowship in that society.

PART V

THE FATEFUL THIRD VOYAGE

«The man who wants to lead the orchestra must turn his back on the crowd.»

James Cook

～

For years, navigators had searched for what they termed the *"Northwest Passage."* It was a search for a waterway that led from the Atlantic to the Pacific Ocean without having to make the very long journey around Cape Horn in South America or the Cape of Good Hope in Africa. It would take months off the journey and expose crews to the dangers of the high seas. The objective, of course, was a productive journey from England to the riches from the lands of the Pacific Ocean, then to Asia and China. In 1776, when this topic arose at the Royal Society, it was intended that the goal of the trip was to be kept a secret. As a cover story, they decided to announce that the purpose of the trip was to return an indigent to the Friendly Islands. His name was *Omai*, and he had escaped death at the hands of the people of Borabora in the South Pacific. *Omai* had been given refuge in England for a year. The man became very famous there, gave lectures, and the British loved him.

~

At the meeting of the Royal Society, Cook leaped at the chance to lead the expedition. He asked for the approval of *Sir Hugh Palliser*, *Philip Stephens*, and *John Montagu* – the Earl of Sandwich. They were thrilled to have Captain Cook himself leading such a voyage. For the journey, the *Resolution* was again to be sailed by Cook, and his sister ship was to be the *Discovery*, commanded by Cook's former officer, Charles Clerke.

~

The *Resolution* had aboard livestock, kersey jackets and waistcoats, pairs of drawers, T-shirts, drawers, trousers, stockings, Dutch caps, woolen caps and plenty of shoes. They remembered very vividly the cold they endured on prior voyages without being well-stocked with warm clothes. The secretive *Northwest Passage* was believed to be well north of the United States, or near Alaska, and would most like be as frigid as Antarctica which they had grimly weathered. This time, they carried top senior officers with a lot of experience. *James King* who had commanded the *HMS Dolphin*, now commanded Cook's sister ship, the *Discovery*. The master of the *Discovery*, *William Bligh*, was the man whom history knew well later on for the mutiny on his ship, the *Bounty*. Captain Cook had aboard several midshipmen, the two cooks, of course, and **William Anderson** the surgeon and part-time botanist. A painter, *John Webber*, was also aboard as well as **William Bayley**, an astronomer. Those and the inked pictures Cook had obtained from the artist assigned to the Antarctic trip contributed a great deal to enhance the written descriptions Cook provided in his meticulous logs. Cook and Clerke left the seaport of Portsmouth in 1776, almost four years after their Second Voyage.

∿

Captain Cook and Commander Clerke pulled into the traditional meeting place in New Zealand, Queen Charlotte's Bay. The island people were again there, and were somewhat apprehensive due to the occurrence of the **Grass Cove Massacre** and feared revenge due to the consumption of the human flesh from Furneaux' crew from the **Adventure**. No expected revenge was forthcoming from Cook, although he was aware of the incident, but didn't know what triggered it. The native people were frightened people who felt they were simply defending themselves in the only way they knew how. The two crews rested at Queen Charlotte's Bay for two weeks to top off their supplies and provisions for the longer voyage ahead. On this third voyage, the men noted that the indigenous peoples sometimes built temporary shelters during the bitter weather. The *hippah* was made up of a mound of mud bricks thoroughly lined with twigs, fronds, and saplings tied up to compose a roof in which they dwelled until the bad weather, rain, and hail passed over.

~

Right nearby were the Friendly Islands they'd seen on the *Second Voyage*, and that's where they let off their passenger, *Omai*, who returned home again finally to rejoin his people and share stories about the British Isles.

THE HUMAN SACRIFICE

～

In 1777, Captain Cook and Captain Clerke reached the island of "*Otaheite*," today known as *Tahiti*. They then took the opportunity to remind the men to moderate their intake of the alcoholic drinks they had on board because the climate would be very cold when they sailed northward toward Alaska. The men would need the liquor for warmth.

～

WHILE THERE, Cook, as the head captain, was invited to visit the chief, *Otoo*. Because the cattle he had aboard had reproduced, he donated some of them to the people – a turkey, ducks, geese, pigs and a few other of the smaller animals. Neither Cook nor any of his men understood the language very well, but they were able to communicate to a limited extent.

∽

DURING HIS STAY, the men from the *Resolution* and *Discovery* were invited to view a human sacrifice. Cook's journals related that he wasn't sure if the victim was a criminal or a selected person for such a ritual. The people of this island insisted that everyone who was a member of their clan worked and contributed to society. Otherwise, the chief and elders rejected them, and they might be sacrificed. Cook said that the victims are never made aware of their fates ahead of time. They are beaten and killed with clubs and stones. It was done with a degree of solemnity, so there seemed to be some religious aspect to it. The victim, according to Cook, appeared to be from a lower class.

∽

THE CEREMONY LASTED for two days. The priests of the tribe first took bundles of feathers and laid them upon stones. Then they offered up prayers. A dog was then sacrificed, and the canine's entrails were cooked and eaten. The dog's heart and kidneys likewise were consumed. Then the human victim was slaughtered and buried in the grave prepared for him. Cook didn't go into a lot of specific details about the ceremony, though he did say that he found the practice abhorrent.

HAWAII: FIRST VISIT

☙

I n 1778, Captains Cook and Clerke stopped at *Kauai*, one of smaller of the Hawaiian Islands. He landed at *Waimea Bay*, and they sailed around the island chain. He was met by groups of the indigent people cheering and waving white cloths.

☙

AFTER THEIR TRADING and haggling ended, Cook and Clerke then turned northeast to explore the western coast of North America in search for a possible entry to the Northwest Passage. This area composed what is now the Southern bulk of California as it is today. In the 18th century, Spain had possession of this colony.

SEARCH FOR THE NORTHWEST PASSAGE

～

The *Resolution* then proceeded up the western coast of the Northwest Territory, followed by the *Discovery*. That territory is currently called Oregon, one of the states of the United States. There they came across massive basalt rocks protruding out into the North Pacific and called it *Cape Foulweather*, as it was always damp and rainy there. Of it, Captain Cook wrote,

> *"The land appeared to be of moderate height,*
> *diversified with hills and valleys and almost*
> *everywhere covered with wood. At the*
> *northern extreme, the land formed a point*
> *which I called Cape Foulweather from the*
> *very bad weather we soon after met with."*

From there, the ships continued north along the west coast and into the territory today known as Vancouver in Canada. They landed on an inlet called "*King George's*

Sound." It was also called "*Nootka*" – a name which was a mistranslation of a native word. The people there are the ancestors of the Inuit Indians or more commonly called "*Eskimos*." Back then, they were called the *Yuquot* people.

∼

THOSE PEOPLE WANTED many more valuable elements for the trade than most indigent tribes, but they had luxurious furs to trade in exchange. No more would the trinkets they had given other tribal nations quite do. The *Yuquot* people wanted pewter, lead, and other metal objects. These people were more domineering as they had to lead the negotiations rather than the other way around. Their chief, though, gave Cook a beaver coat in exchange for a broadsword and hilt. Cook felt that it was a truly fair exchange.

∼

UPON LEAVING *NOOTKA*, they headed for the *Bering Strait*, and the archipelago that stretched westward off the coast of current-day Alaska now called the *Aleutian Islands*. Cook and his men discovered that the Bering Strait was literally impassable, covered with solid ice that allowed for no inlet eastward.

∼

ONE OF THE Aleutian Islands was *Oonalashka*, also called "*Unalaska*" by some. The *Eskimos* in that region lived in long houses made of driftwood over the ribbed framework of the roof and the walls. Branches and saplings are twisted together, then held with mud. Ones that were made well were dry inside. There was a hole in the roof to allow for entry on a tall mounted log and another hole to allow the smoke from campfires to escape. These long houses were like open apartment buildings which clans shared. A primitive sanitation system was also part of the structure, although its effectiveness wasn't superb.

∼

THE PEOPLE of the Aleutian Islands were friendly and had also been accustomed to doing business with the Russians who crossed the Bering Strait when it was frozen over. Cook and his men described the inhabitants as being "*peaceable and inoffensive people*." They seemed to have integrity and were honorable.

FRUSTRATION

≈

F ruitlessly, Cook and Clerke spent months searching for a waterway eastward. It was getting late in the year, and much of the water was iced-over. Also, their supplies were running very low. Cook then required his men to consume walrus meat, which they found to be practically unpalatable. They did come across a field of ice strewn with seahorses, and they consumed those in great numbers. Like the walrus meat, they were salty and tasted bitter and chewy. It was also impossible to fish.

≈

CAPTAIN COOK himself was becoming ill. Likewise, Clerke himself had been developing tuberculosis and was showing all the initial symptoms of it. Cook was depressed and frustrated because his mission was unsuccessful and was argumentative. He was the kind of man unaccustomed to failure. The sailors also started to feel angry and confused about

Cook's erratic behavior and his insistence that they ate poor food when there were still some stores of food below deck.

~

BECAUSE THEY WERE in dire need of food and provisions, the crew turned southward to return to Hawaii. This time, the two ships visited the largest of the Hawaiian Islands called *Oahu*. Much time had passed during their exploration of the *Northern Pacific*, and it was now the year 1779.

~

UNBEKNOWNST TO HIM, Captain Cook and Clerke arrived at the Hawaiian Islands during the festival of *Makahiki* which celebrates *Lono*, the god of fertility, agriculture, and renewal. It is a time associated with tropical rainfalls which required strict observance of a time of rest interspersed with numerous ceremonies and many dance festivals. Violators who work during that time were punished.

~

SOME HISTORIANS INDICATE that Cook's return was perceived by the native people as a reincarnation of the god, *Lono*. However, *Lono* was not only a god of renewal but seen as a deity that instills much fear. One of the reasons for that was the fact that this season was marked by storms, fierce rainstorms, and gales.

HAWAII: SECOND VISIT

≈

B ecause he was the chief captain, Cook was escorted to a sacred structure made of thatch and wood. It was called *"Harre-no-Orono,"* or the house of the *Orono*, a god represented by a large sacred idol. The idol was wrapped in red and accompanied by twelve priests. A parade formed, and the people of the village slaughtered a pig which they carried to the fire. When the hog arrived, the people prostrated themselves before the god, *Orono*, and prayers were conducted by the high priest. The pig was then slowly roasted in the embers. Music and dancing preceded the feast, taking place while the hog was cooking. This ceremony was similar to the *luau.*

≈

COOK AND CLERKE'S seamen conducted a great deal of trade with the indigent people there, but a lot of stealing took place. The seamen had to be careful with their equipment as well as their goods since the natives would take anything that looked like it could be converted into something practical for themselves. The main island of "*Owyee*" or *Oahu* was much larger, but also volcanic. It had a very hot but pleasant climate cooled by the trade winds. Cook spend several weeks exploring the entire island. The people approached the **Resolution** and **Discovery** in their canoes, regularly engaging in trade. They also had Cook sample sugar-cane and a delightful beer made from it. Once he added a few hops, it improved its taste significantly.

～

Huge crowds of the islanders followed him about as he explored the island. He later disembarked and sailed into an adjacent bay called "**Kealakekua**." On occasion, the crowds who greeted the men became rowdy and the high priest, **Koa**, had to calm them. In the meantime, Cook and his men repaired the riggings on the ships and caulked the worn timbers on the side of his vessel. Also, they kept watching against the stealing that went on continuously. Cook's men themselves also stole items – namely wood from protective fencing that surrounded their sacred burial grounds – an unwise decision. They desperately needed a lot of timber for his vessels because it was badly worn or broken from collisions with the sea ice.

～

The king of Hawaii, **King Kalei'opu**, traditionally didn't make an appearance until a few days had passed. The king was accompanied by the chief priests and solemnly approached Captain Cook, offering him his cloak. Then everyone feasted.

~

AFTER THE SHIPS were repaired and refitted, they stocked up on food and provisions and set sail for home. As they pulled out of the harbor and bid their farewells, an enormous storm blew in without warning. Huge swells arose out of the sea in the, and the ships were shoved much further inland. Lightning, thunder, and rain were all about them. Both ships were caught in the middle of this horrid gale. Suddenly, the rope from the mainsail of the **Resolution** gave way, and it tore in two. Then the top sails likewise broke. Men who were ashore came over when the winds died, and the storm abated. Then some of Clerke's men hopped aboard and steadied what masts they could save. Canoes filled with the inhabitants came over in great numbers, offering their help. However, one of them stole the rudder from the **Resolution**. Cook had his men fire muskets at the offender's ship, after which the thieves took off.

~

THE HAWAIIANS who were ashore asked if they could come aboard and sleep on the ship for the night. The two captains permit them to do so to show good will. However, that was a mistake. Many more items were missing in the morning.

~

THE SEAMEN WERE able to rebuild the rudder, put it into place, but because of the great storm; they moved into *Kealakekua Bay*, which might offer more protection.

WELCOME WORN

~

Accoding to one of the passengers, *John Ledyard*, an American,

> "*Our return to this bay was as disagreeable to us as it was to the inhabitants, for we were reciprocally tired of each other. It was also equally evident from the looks of the natives as well as every other appearance that our friendship was at an end.*"

When they were anchored in the bay, one of the two long boats called cutters from the *Discovery* went missing. Cutters were used to shuttle the sailors back and forth to and from the shores. Every one of these long boats was crucial, and they couldn't continue with its loss. Clerke informed Cook, and Cook – already short-tempered – was furious. He then made a series of foolish moves. He issued orders that some of his marines go after the natives' canoes in search of it. They left in another cutter, armed with muskets and eventually discovered that the suspect's name was *Kariopoo*. *Kariopoo* had landed at a local island, and the men went searching for him and the boat. The people of the islands seemed very concerned and willing to help. However, Cook felt that they were insincere, just giving him lip service. His anger then reached the boiling point, and Cook hatched a devious plan to kidnap the chief and hold him until the cutter was returned.

THE DEATH OF CAPTAIN COOK

~

Cook sent messengers into the king's house, demanding the meeting with *King Kalei'opu*. Cook then noted that some of the men had armed themselves with spears, daggers, and clubs. The women and children who were usually outside washing the clothes on the beach had left the area and returned to their homes. Cook and his marines resolutely proceeded into the king's enclosure and forced him out. The minor Hawaiian chieftains, *Kana'ina*, *Keawe'opala*, and their attendant, *Nuaa*, closely followed to act as bodyguards. Behind them was one of the king's wives, *Kahekapolei*, who cried out after Cook begging that her husband be released. When the king reached the beach, he sat down and refused to cooperate. An old kahuna (elderly priest) offered Cook and the others a coconut and accompanied that with chanting meant to prevent an altercation. That was to no avail, as Cook and his men shoved them aside. By now the shore was loaded with armed Hawaiians. Cook screamed at the king, telling him to stand up, but his bodyguards surrounded him. One of them, *Kana'ina* approached Cook, but Cook hit him on the head with the broad side of his sword. *Kana'ina* then grabbed Cook and lifted him high into the air, after which he slammed him down into the sand face-forward. As Cook was rising from his prone position, the bodyguard, *Nuaa*, stabbed him with his dagger. Cook fell to the ground, and his head was half-submerged in the bloody sand. He was dead. It was February 14, 1779.

~

THE SAILORS and the natives then started slaughtering each other. Four of the marines were killed, and cannons were fired from the *Resolution* and *Discovery*. In revenge, Bligh and his men attacked the village. Captain Clerke then demanded the return of Cook's remains. After some heated discussions, Cook's remains were returned. Clerke reverently took the body of James Cook out to sea where he was given a sailor's burial.

~

CLERKE, who had been suffering from tuberculosis, expired before reaching England. His first mate, *John Gore*, assumed command of the *Discovery* and *James King* took over the *Resolution*. Grief and mourning spread across England upon the news of Cook and Clerke's deaths. There was no excitement and no parades when the *Resolution* and the *Discovery* docked at Portsmouth, England on October 4, 1780. The tragedy hovered over Great Britain, but the people read

Cook's journals and all the literature about the courage of these men and greatest voyages ever taken during the 18th century.

PART VI

CONCLUSION

~

Captain Cook left behind him a legacy no one else could. For the *Royal Society* in London, he circumnavigated the world twice. Also, Cook charted and drafted the continents. During his lifetime (1728-1779), he saw the wonders of the world and traveled further north and south than any other human being of his time. His journals record his experiences with different cultures, both cordial and hostile. Besides the exploration, Cook took astronomical measurements and observed an eclipse, as well as the phases of Venus and the planets. *James Cook* was one of the most skillful navigators ever known, as he was able to preserve his ship through gales, storms hailstones, the ice of both the Arctic and Antarctica.

PART VII

FURTHER READING

∽

- Beaglehole, J. C. (1992 reprint) *The Life of Captain James Cook.* Stanford University Press.
- Edwards, Philip (ed.) (2003) *James Cook: The Journals.* Penguin Books.
- Kippis, A. (1871) *A Narrative of the Voyages Round the World Performed by Captain James Cook.* Claxton, Remsen & Halefinger.

Copyright © 2019 by Kolme Korkeudet Oy

All rights reserved.

No part of this book may be reproduced in any form or by any electronic or mechanical means, including information storage and retrieval systems, without written permission from the author, except for the use of brief quotations in a book review.

YOUR FREE EBOOK!

As a way of saying thank you for reading our book, we're offering you a free copy of the below eBook.

Happy Reading!

GO WWW.THEHISTORYHOUR.COM/CLEO/

Printed in Great Britain
by Amazon

80715752R00058